LOVE FROM KEW

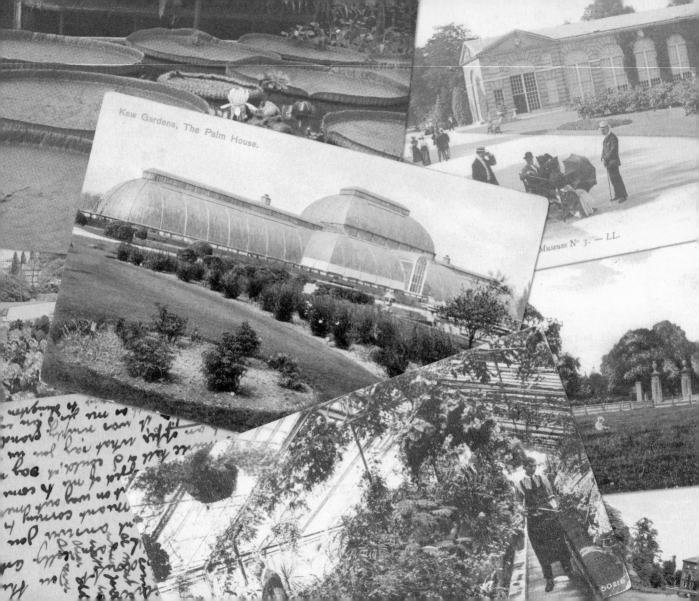

Kew Gardens, The Palm House.

Museum No 3. — LL.

50216

LOVE FROM KEW

A postcard scrapbook

SOPHIE SHILLITO

Kew Publishing
Royal Botanic Gardens, Kew

First published in 2020
Royal Botanic Gardens, Kew,
Richmond, Surrey, TW9 3AB, UK
www.kew.org

ISBN 978 1 84246 732 9

Distributed on behalf of the Royal Botanic Gardens, Kew in North America by the University of Chicago Press, 1427 East 60th St, Chicago, IL 60637, USA.

British Library Cataloguing in Publication Data
A catalogue record for this book is available from the British Library

Design and page layout: Ocky Murray
Image work: Christine Beard
Project manager: Lydia White
Production manager: Jo Pillai
Copy-editing: Michelle Payne
Proofreading: Gina Fullerlove

Printed and bound by in Italy by L.E.G.O. S.p.A.

For information or to purchase all Kew titles please visit shop.kew.org/kewbooksonline or email publishing@kew.org

Kew's mission is to be the global resource in plant and fungal knowledge and the world's leading botanic garden.

Kew receives approximately one third of its funding from Government through the Department for Environment, Food and Rural Affairs (Defra). All other funding needed to support Kew's vital work comes from members, foundations, donors and commercial activities, including book sales.

MIX
Paper from responsible sources
FSC® C023419

For Mum, who taught me to love words and plants.

CONTENTS

PREFACE

PREFACE

I want you to experience the postcards in this book just as
I first did, in a haphazard, unordered, accidental way – as
though you are rummaging through a shoebox full of them, or
finding a crate of them in an attic, turning the gluey pages of a
scrapbook full of precious clippings, stumbling across the past
in the middle of the present. Through the prism of the cards,
I hope that you will discover the wonder of Kew, just like the
postcard writers who visited all those years ago.

   I want to share the curious, charming beauty of the cards –
the buildings and the plants, the visitors captured mid-stride.
The colour of the stamps, still vibrant after all these years, the
faces of the monarchs in reds and greens, pasted down with
spit – the very DNA of the sender stuck to them. The pure
aesthetic pleasure of looking at the postmarks and timestamps,
the blotted hooks and loops of a fountain-penned hand giving
away details of the writer's personality in a way that only
handwriting can – the words falling from the end of their arm
in a wobbly script, or a confident hand printing firm letters.

The best thing about the cards though, I think, is not the pictures (although some of them are very attractive and interesting), or the handwriting or the postmarks and stamps, but the stories. I am fascinated by the social anthropology, the microhistory – by the scrawled memos that I have to sound out loud to decipher, turning my head sideways and squinting like a fox in the undergrowth, at the puzzle of moments and the fabric of everyday life captured in inky missives. Here is humanity, in all shades and flavours, in messages which are poetic, cryptic, funny and poignant. Postcards are tiny windows into lives.

Kew has bewitched me with its cornucopia of stories and the narratives of the postcard writers preoccupy me just as much. In this book I have tried to tell both. This is not a book about postcards, but a book about people and their relationship with Kew, and about their relationships with each other told through the prism of the Gardens. Through my writing I have tried to weave their stories together with Kew's secrets – their lives are entwined with the Gardens, tangled as a vine.

My overactive imagination has helped me to fill in the gaps. I have taken cuttings of information and propagated them into postcard prose. Lost in an enchanted thicket of vignettes deeper and muskier than the woods down by Queen Charlotte's Cottage, I have begun to forget what is real and what I have imagined. I have taken fact and woven it into folklore. If you like, you can think of what I have made as an alternative, imaginative guidebook, a love song to Kew, as fabulous as the follies sprinkled through the Gardens.

ROYAL MAIL

POSTAGE PAID

GREAT BRITAIN

W7047

PALACE

INTRODUCTION

INTRODUCTION

A garden full of stories

On a map of greater London, Kew is a green oblong sandwiched between the Thames and the A307, resembling any other natural space in the vicinity: Syon Park across the river or Richmond Park to the south. A verdant smudge between the centre of town and the airport − you could quite easily be forgiven for thinking it's just a garden.

And indeed, Kew is a garden − a beautiful one. Every year, people like you and me visit in their millions to walk under the shade of the towering redwoods, to marvel at rare cycads, to picnic on Syon Vista. Together with hordes of day trippers, global tourists and locals, I have delighted in the glorious flowerbeds and steamy glasshouses. In summer I have sought sanctuary from the heat in hallowed architectural follies and in winter I have wrapped up warm and taken long walks in the arboretum, enjoying the Gardens for their own sake.

But Kew's landscape is just the beginning. As it became more familiar to me, I suspected there might be more to know than what was immediately visible. I suspected that Kew might be a garden full of stories and secrets to be discovered – they mumbled to me through the turf on Kew Palace lawn, and whispered on the wind that blows through the prickly bushes on Holly Walk. History is fluid, and won't stay buried – it rises up through the soil. The past writes its truth on the landscape and is not so very far away, after all.

I began to understand – in a way I could only understand by walking through the landscape, rather than looking at the map – that Kew is a trove of wonder, a whole kingdom in its own right. Kew got under my skin and enchanted me. I was hooked, compulsively and zealously. I was in love. I wanted to know everything I could, about the plants and how they came to the Gardens, about the land itself, but more importantly, or perhaps *most* importantly, I wanted to know about the people. Not the grand people with grand names from Kew's history, like Princess Augusta, Decimus Burton or Sir Joseph Banks – I could read about them in books (and so can you, if you want). No, I wanted to know about people I could relate to. People just like me – and maybe just like you – who visited Kew and loved it, too.

Where would I look for them? Not in a book. Where else? Perhaps a museum. Museums are excellent places. At Kew, over time, there have been several. In a museum, you can learn about history by looking at objects that have existed in two places: the past and the present. In the past, the objects were useful and now they are more decorative. For example: a spade which was once a helpful tool for digging holes to plant trees is

now venerated, trapped in a sterile glass case. You can look at
it of course and read the label, but that's exactly my point. You
can look. With your eyes, only. I wanted to look with my hands.
To hold history in them and turn it over. To smell it. I find I
can discover much more that way.

So I began searching elsewhere for the people who visited
Kew. I wanted to know about their lives, their everyday. What
brought them to the Gardens and what did they see? Were they
like me, or different? What were they thinking and feeling?

I thought carefully about how we express our thoughts
and feelings today, how we communicate: often through text
message or social media. I thought back to my childhood
(before either of these things existed) and closed my eyes to
recall what we used to do. I remembered the picture postcards
my grandmother would send me, written in her shaky hand,
about the cake she had baked, farm animals she had seen,
sunny days at the beach. Maybe postcards were my portal to
people from the past.

I bought one Kew postcard, then another, and soon,
almost without realising, I had a small collection of musty old
cardboard rectangles . . . and a postbag full of stories.

Postcards from the past

While various forms of printed pictorial stationery – such as
greetings cards and printed envelopes – had been available
from the 1840s onwards, it was not until 1861 that a printer
from Philadelphia called John Charlton obtained the copyright
for the first privately printed postcard. For reasons lost to
history, Charlton then transferred the copyright to his friend

These green sugar-
paper envelopes
contain sets of Kew
postcards from 1925.

and fellow Philadelphian, Hymen Lipman, and plain postcards bearing the text *Lipman's Postal Cards* were printed; however, probably owing to the US Civil War they were largely forgotten about and the first record of one of Lipman's cards actually being posted does not appear until nine years later. Lipman, by the way, is also credited with inventing something you might need to write your postcard (and delete any mistakes): a pencil with an eraser on the end.

Meanwhile, across the sea in Austria, Dr Emanuel Herrmann – a political economics professor who liked to collect folk songs – sat at his desk and wrote an article entitled *Uber Eine Neue Art Der Korrespondenz Mittels Der Post* (A New Way Of Correspondence By Post), which promoted the use of postcards instead of costly-to-send letters. Dr Herrmann's article was published in January 1869 in the Viennese newspaper *Neue Freie Presse* and on 1 October that same year – just over 150 years ago – the Austrian Postal Authority introduced what are widely accepted to be the world's first official postcards: fawn-coloured oblongs bearing the words *Correspondenz-Karte* and the mutton-chopped face of Emperor Franz Joseph on a butter yellow two-Kreutzer stamp. Post Office regulations concerning the cards included the need for the sender to write both the recipient's name legibly and durably, using ink, pencil or coloured pencil, and that no obscenities or libellous remarks should be written on the cards. Approximately one million cards a month were sold before the end of the year, providing a boon to the Post Office coffers.

The British Post Office saw the success of its south-eastern neighbour and quickly followed suit, releasing its first official plain postcards exactly one year later, on 1 October 1870 – a

A set of six Kew
postcards featuring
watercolours of the
Palm House, Rock
Garden, Japanese
Gateway and Lake.

halfpenny violet Victoria in the top right hand corner stared intently at the inscription *The Address Only To Be Written On This Side*. During October, exhausted postmen pushed one and a half million of the new cards through letterboxes across the land, and within a year 75 million cards had been written and sent.

The government maintained their monopoly on postcards until 1894, the same year that the first British picture postcards were issued – showing enticing views of Edinburgh and Leicester – and postcards became even more popular, with their 'Golden Age', typically considered to be the first

20 years or so of the 1900s, until calling friends and family on the telephone became more popular and taking pictures of interesting scenes with one's own camera became more affordable. Indeed, the majority of the postcards in this book were sent during this time. Just as texting and tweeting have grown rapidly in popularity today, postcard-passion swept the nation, peaking in 1913–14, when a record 926 million cards were sent – 2.5 million a day. To keep up, the Post Office managed six deliveries a day in towns and cities and 12 in the centre of London – postcards sent first thing in the morning could reach their recipient by lunchtime.

A result of so many postcards being sent is that it became very easy to amass them simply by waiting in for the postman, and collecting them quickly became a craze. Postcard collectors were labelled variously deltiologist (in the US), cartologist (UK) and cartophilist (France). Specialist magazines such as the monthly *Picture Postcard* circulated, there were fairs and exhibitions of postcards and collecting clubs were established, where cards were admired and traded. A couple of the postcards in this book tell collecting stories: 'We were getting cards at Kew today and I send this for your collection'; 'Will you ask Bert to get another album for p.p.c's [picture post cards] like he had, red if possible.' Now, over 100 years later, I have collected the very same cards again.

Kewensia

As part of the research for this book I delved deep into Kew's own archive. In the Library – part of the Herbarium complex on Kew Green – among the books about natural medicines

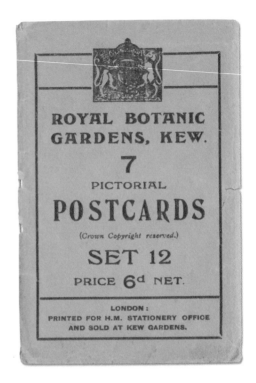

ROYAL BOTANIC GARDENS, KEW.
7
PICTORIAL
POSTCARDS
(*Crown Copyright reserved.*)
SET 12
PRICE 6d NET.

LONDON:
PRINTED FOR H.M. STATIONERY OFFICE
AND SOLD AT KEW GARDENS.

A set of seven Kew postcards of Orchids, showing the Lady's Slipper Orchid and the Blue Orchid.

from New Guinea and records of collecting trips to Thailand and Colombia is a special section dedicated to Kew's own self-interest, labelled *Kewensia*. Shelves filled with all things Kew, collected by former members of staff who felt the organisation's history when it was their present and had the presence of mind to hoard their everyday documents, knowing they might be interesting one day. Thank goodness they did – because of them, there is a vault rich in information, and some of this is about postcards.

Nose-deep in the archive, I learnt, for example, that in May 1926 Kew postcards were available in both 'black-and-white and in natural colours' and could be purchased at the Publications Kiosk behind Museum III (now the Orangery) or by post from the curator, either as single cards for a penny, or as sets in a 'descriptive folder', with coloured cards costing twice as much as the monochrome versions.

A Kew catalogue from March 1947 offers postcard sets for sale. Subjects for each set were numbered and themed – for example: *Stove and Greenhouse Plants, Rhododendrons, Orchids. A Selection of Outdoor Subjects* included *The Iris Garden* (now the Grass Garden) and *The Blue Border in Cambridge Cottage Garden*, while *A Selection of Indoor Subjects* featured *The Sherman Hoyt Cactus House* and *The Alpine House*. One of the numbered postcards in my collection was sent to Enid – 'a No 6 as promised'.

Perhaps as a way of enticing visitors to buy multiple postcards, the Kew catalogue states that the sets 'in some instances include cards which are no longer available singly.' Also interesting to note is the proclamation that 'no new postcards have been printed since the war' but that 'the postcards listed are still available and may be purchased by post from the Curator.'

A postbag full of secrets

When postcards were first introduced, people worried that it would be incredibly easy for their private messages to be read by the postman, by the butler who brought them on a silver tray, by the nosy maid, by a visiting aunt who picked them up from the mantle to peer at them through her lorgnette. There have been times when I have read the messages myself and felt as though I am doing something I shouldn't – trespassing, pushing open a creaky old door in a wall and peeping into a secret garden, eavesdropping outside the greenhouse. However, I am yet to uncover anything extraordinarily scandalous.

Indeed, the postcards tell stories about the quotidian. About Christmases and birthdays, snow and heatwaves. About exams and school trips, illness, summer holidays, bombs and dancing and love.

Trends and changes

The postcards reveal social trends – lots of them are written by women and lots of the women are called Nellie. Visitors and gardeners captured in the pictures provide a good record of fashion – from hoop skirts to orange sheepskin coats. There are cards written by people from different financial backgrounds – Auntie Jojo can afford four weeks in Spain followed by a sojourn at the beautiful seaside resort

of La Baule on France's Côte d'Amour, but Marianne cannot afford a birthday present for her cousin Annie. Whatever their names or outfits or however many coins line their pockets, the abiding resonance I draw from the cards is that the writers all loved Kew.

Changes in the Gardens are reflected, too – an early postcard shows flat-capped gardeners sweeping the gravel on Pagoda Vista, now a grassy swathe; the Number Four Greenhouse has long been demolished. The cards also show that it is not just the landscape that has changed, but also the way it is referred to – Kew's 1947 postcard catalogue lists cards with pictures of places that no longer exist, or rather, that have been erased and written over: *Oak Avenue*, the *Himalayan House* and *Nepenthes House*. Spellings have also changed: Syon Vista was once Sion.

Questions and answers

Part of the beauty of the postcards is that they provide more questions than answers. What about that particular view of Kew appealed to the person who sent it? Did they feel the heat of the Palm House depicted on the front, or sit on a bench outside the Orangery? What was their day like? Who did they visit with? Just as the wonder of Kew is the wonder of the partially unknown and unknowable – the hidden things in the woods, the forgotten heat of summers past, the secrets under the soil – the delight of these postcards for me, as I hope it will be for you, are the half-stories, the hints and glimpses of lives lived. What they have told me about Kew and about life is fascinating and that is why I want to share it here.

Many of the postcards in this book feature vibrantly coloured stamps depicting the monarch of the time.

Because of the postcards, Kew's familiar landscape has changed again for me and revealed itself to be a palimpsest. The Lake is no longer just the Lake, but the place where a couple sat in the shade of a pine tree. The Temperate House has become the place where a man had sheltered from the rain and as I walk through the Rhododendron Dell I think of the woman who had applied for a passport to return to New York, wondering if she made it and what became of her.

Private detective

I engaged in what I began to think of as 'method writing'. I wanted to feel what the postcard writers felt, to experience their Kew, to inhabit them, to *become* them. I stood underneath the ginkgo tree and peeled a wet leaf off the path, pressing it in my notebook, in an attempt to understand Gladys. On a day filled with drizzle I searched all over the Gardens for the Corsican pine and lent against it, feeling its peeling bark. I collected its cones and watched the resin seep from them onto my desk, so I could feel closer to Dorothy.

I became a detective and also a little obsessive. Armed with addresses (which have been blurred out in this book to protect the privacy of current residents) and Google Street View, I stared through my computer screen at the houses where the postcards were sent, zooming in on the letterboxes to see the slots they dropped through, imagining the *splap* as they fell onto mats and the race of little feet as children skittered down hallways to see what the postman had brought them. On a weekend in Paris, I sought out the Boulevard de Clichy, so I could gaze up at the proud townhouse where Mademoiselle Philippe once lived.

A 'Kew Gardens' postmark from 1909.

I love holding the postcards in my hands, because Claude and Sonny and Kath held them in theirs as they walked to the postbox on their way to the train, or to school, or en route to the shops. I feel sometimes, by holding the postcards, that I am holding their hands. I feel, sometimes, as though they were writing to me.

Return to sender

In a way, the postcards in this book have come full circle, returning to Kew with new tales to tell about where they have been in the meantime. Recently, I visited Kew's gift shop and watched visitors choosing postcards to send, sharing their love of Kew in turn. Time moves forward, but history repeats.

I hope you will enjoy looking at the postcards as much as I do and that my imagination makes you curious. I hope that you will visit Kew and buy a postcard and send it to a loved one, sharing a thought about your day, so that maybe in the future someone else will read it too, and wonder, and wish they were there. Which picture would you choose, where would you send it, and what would you say?

Sophie Shillito
November 2020

THE POSTCARDS

The muddy silt bottom lets him go. He feels the moment the current catches him and he is afloat, graceful and buoyant. He smiles as the river plays with the boat and cuts downstream as sharply as pinking scissors slicing through felt. A spider catches its silk on the gunnels, casting a gossamer line, hitching a ride. Waterweed anchored to the riverbed grazes the hull, making a hissing noise – a needle on a record. Willows cry their tears. Cow parsley foams cream umbrellas on the banks. Feathery damselflies dance in the air, flitting around the prow, kissing it, leading him to the garden.

Printed text on the reverse of this postcard reads: 'The Thames and Isleworth, from the end of Sion Vista. This View was painted by both Turner and Richard Wilson early in the nineteenth century.' 'Sion' is now spelled Syon.

The earth turns, first light stains the horizon. A finch darts over the garden wall, over the Pagoda, over the redwoods and the Berberis Dell. Mossy trees wearing bracket fungus wave evergreen fingers towards the Temperate House, where tubs froth cadmium geraniums. Lines of red oak and Norway spruce border wide vistas which lead through shrubberies to the rye-thatched cottage in the woods. Rotten branches litter the floor, the air smells of soil.

This postcard shows a view taken from the Pagoda, looking up Pagoda Vista towards the Temperate House and the Pavilion. The Palm House can be seen in the distance.

Jo loses her sense of time and direction in the trees at the bottom of the garden. Dew leaks through a hole in her boot, dampening her sock. She finds her childhood home in a clearing. Is that her mother waving from the upstairs window? She remembers the lilac that grew by the door – how purple it smelled after the rain.

'What does the picture overleaf remind you of? Can you smell the lilac blossom!!! Love Jo'

This postcard of Queen Charlotte's Cottage was sent on May Day in 1945 from Kingston-upon-Thames to Harrogate. The building is a cottage orné, a picturesque, rustic cottage built to decorate a landscape, rather than to live in.

After dinner she clears the crockery and shakes the cloth into the back garden where the sun is setting. She lights the lamp and walks into the lane where a blackbird sings its evening song. Tonight her girl is coming home.

I have been in these gardens and they are beautiful. M.M.M.

The Palm House Kew Gardens

E. Whiteley, Kew Gardens

POST CARD

THIS SPACE AS WELL AS THE BACK, MAY NOW BE USED FOR COMMUNICATION. (Post Office Regulation).

THE ADDRESS ONLY TO BE WRITTEN HERE.

'I have been in these gardens and they are beautiful.'

'Dear Mother, we shall start after tea this evening about six, so be on the lookout for us.'

This postcard shows the Palm House and was sent in August 1905 to a Norfolk village.

Miss Stephens is as pale as the gossamer dress she wears to church on Whit Sunday, and just as delicate. She thinks she might blow away in a puff of wind, like a dandelion seed.

'Dear Miss Stephens, I was sorry not to see you at class but you did not miss so much as sometimes. I do hope your head is better and you had a nice time at Whitsuntide.'

This postcard sent from Putney in the summer of 1905 is labelled 'Entrance to Kew Gardens' and shows Elizabeth Gate from Kew Green. Several properties on Kew Green are connected with the Gardens, including number 49, the director's official residence, where Kew's first two directors, William Hooker and his son Joseph Hooker, once lived. They are both buried in the churchyard of St Anne's, also on Kew Green.

London. Entrance to Kew Gardens.

THE ROCKERY,
KEW GARDENS.

Water sparkles through rills, pink sedum grips the granite. He watches fleshy lambs ears lollop down terraces, under pines dripping sweet golden resin. Campanula sounds tiny chimes, ringing with the song of a mountain – a skylark, a stream, a thyme-scented breeze.

This postcard showing a Kew Constable in the Rock Garden was sent in March 1907 from Forest Hill to Brighton. Kew Constabulary was established in 1845 and comprised Metropolitan policemen and, from the mid-1850s, army pensioners who had fought in the Crimean War.

Little fingers trace the writing on the sign outside the Alpine House – words are sounded out in a whisper. *These plants live in a world of ice and snow, high up in the mountains, where summers are short and winters are long.* Where, above the tree line, a horned animal with cloven hooves stands proudly on a ledge. Crickets rub their legs together in the tussocks sprouting round the shore of an emerald tarn. Silvery plants clamber over rocks which bleat with echoes, thriving in the harsh conditions they were made for, pushing roots through the arid scree, taking hold, making a life.

The original Alpine House was built in 1887 within the T-Range glasshouse complex and demolished in 1983 to make room for the Princess of Wales Conservatory. In 1981 Kew opened a new Alpine House shaped like a pyramid, but this was dismantled in 2004 to accommodate an extension to the Jodrell Laboratory. Construction of the modern Davies Alpine House began in the same year.

Tiny white crocus peep delicate faces through the snow under the stone pine, sticking out saffron tongues.

'What rough weather we are having, strong winds and snow.'

This postcard shows the stone pine. Before being planted in the Gardens in 1846 it was kept in a pot for many years, which caused it to grow an unusual number of branches. It lost a branch in a snowstorm in January 1926 and gradually tilted to one side. A prop was then used to support the tree from the spring of 1941, until another branch was eventually removed in the early part of this century.

She fumbles to pull the string and open the brown paper. Inside is a crocheted white hat. She presses it to her face, remembering the one she knitted her daughter, all those years ago.

13102   The Lake, Kew Gardens.

'Thanks very much for parcel. Very nice indeed. I love it and will write soon. Been here today and had rain and wind and came home in the snow. My feet and hands have been bad lately.'

This postcard was posted in April 1917 from Hammersmith to a village outside Gosport and shows the Lake, which is filled with water from the river Thames.

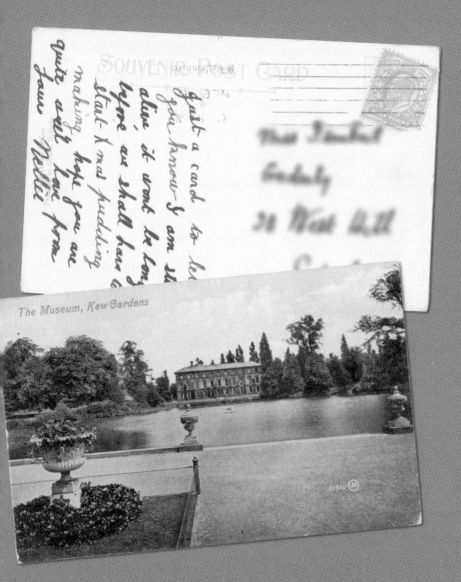

The Museum, Kew Gardens

She makes a special pudding using a recipe her grandmother wrote and decorated with pencil pictures of the ingredients in the margins of her feint ruled cookbook – a basket of eggs, a square pat of butter wrapped in paper. The book is stained from years in the kitchen – sticky candied peel and a wizened currant, squashed between the pages.

'Just a card to let you know I am still alive, it won't be long before we shall have to start Xmas pudding making. Hope you are quite well. Love from your Nellie'

This postcard sent to Sydenham in 1907 shows a view across the Pond to Museum Number One, which was built in 1856 to display plant-based objects such as clothing, tools, medicines and food.

Pinned to the corkboard inside the horticultural school are the monochrome portraits of the class of '78. Young, rag-tag, hopeful, hairy – mugshots from a different time. Years turn – the view changes and stays the same. They become part of the land they take care of. Storms are weathered, harvests gathered, skin becomes leather, browned by 40 summers. They celebrate with cake and tea. Friends, still.

Kew's School of Horticulture (originally a gardener's residence and store, then a museum), where students study for the Kew Diploma, has produced several famous alumni. Photographs of each year's cohort are displayed inside. Many Diploma students continue their careers at Kew.

35930. London. American Gardens. Kew.

In the quiet of the morning, before the gates are open, before the hordes arrive with their baskets and blankets, gardeners make their rounds, sweeping gravel walks, trimming grass moustaches to match their own. Robins pull worms from the edges of lawns – freshly straightened with sharpened spades.

This postcard shows a view down a gravelled Pagoda Vista from the American Garden, which was situated behind the Palm House.

V 4671   KEW GARDENS, ENTRANCE GATES.

Opening time. John unlocks the gate with his heavy brass key.
The eager queue shuffles in – rucksacks, flat caps, sweaty bank
notes, flasks and sandwiches. The turnstiles click. Here they
come, whirling their maps round in the air, trying to make sense
of the shapes and symbols, tracing fingers over contours and
pathways, pointing at drawings of trees.

This postcard shows
Elizabeth Gate. The
turnstiles were introduced
in 1916 when the admission
fee for the Gardens was one
old penny.

She tells Mrs H about the garden. How children wearing only their knickers run through the sprinkler, how the red hot pokers are ablaze. How ices from the cart cost only threepence. They taste so cold.

'We are having a nice time here today. Hope to go and see Mrs H tomorrow, lovely weather here hope it still continues.'

This postcard was sent in August 1951 from Kingston-upon-Thames. The montage of images shows St Anne's Church on Kew Green, the Temple of Aeolus and the Waterlily Pond.

A girl puts her head on the grass by the side of the Pond, pushing on the lawn with her bare toes. Her legs lift, her feet touch the bottom of the sky. She enjoys the feeling of the blood draining downwards into her face. A green-stockinged moorhen creeps away, fluffing a white bustle.

The Palm House Pond lies on the site of Kew's original lake, which was much bigger and had an island that could be reached by the Palladian Bridge, reputedly built in one night.

THE ENTRANCE GATES. KEW GARDENS.   316

People come from the city, in carriages that fill the road, in boats that stop at the water pier. A queen and princess in snow-starched frills, kid boots and cloche bonnets – Sunday best for a royal garden.

Writers of several postcards in this book describe sailing to or from Kew by riverboat. This postcard shows Elizabeth Gate, the closest gate to Kew Pier, where boats moor.

In the cottage, the queen and princesses perch on cushions trimmed with fancy fringes blacker than the queen's eyelashes. The table groans with courses – minced fowls with strong pepper, a brace of partridge, loin of veal cooked in fat and capers, scotch collop in a pastry coffin, pickled roots with butter. Yellow cabbage, apple dumplings, ice fruit, jellies and blancmange. The pretty hens peck it all up, passing dishes to one another across the table in an elaborate game of culinary chess until the queen – plump with more than children – declares check mate by finishing off the rabbit larded with bacon and wiping greasy fingers on her damask-covered knees.

This postcard was sent from Richmond in May 1920 and shows Queen Charlotte's Cottage, thought to have been built in the late 1700s. Queen Charlotte was George III's wife. Members of the Royal Family living in Kew Palace would visit the cottage during walks in the Gardens, and dine there, in an upstairs picnic room.

13098

Kew Gardens and Palace.

Royal treasures hide in a shallow grave, just beneath the turf – a painted teacup, a candlestick, a small glass bottle, its stopper stuck tight. Nails from an ornate screen, a prickle of pins. Families promenade where young princes once played chase-me, trip-trapping on courtyard cobbles, jumping over box hedges planted with nosegays of borage and soft sage.

This postcard was sent in September 1907 and shows Kew Palace, originally built in 1631 for Samuel Fortrey, a wealthy London silk merchant. The most famous resident of the Palace was George III, for whom Kew was a secluded countryside retreat away from the excesses and pressure of London. He lived there with his wife Queen Charlotte and their 15 children.

The people point at the
little flowers, calling them
names: *bell bottles, witches'
thimbles, crow's toes.* They
bend their heads, weeping
sticky sap tears down
their stalks.

**'I thought you might like a p.c.
Have you seen the bluebells
in Kew Gardens yet? I know
Mother and Daddy usually go.'**

This postcard was sent in 1923.
It shows a black-and-white
view of bluebells growing in the
Natural Area. Bluebells have
many common names and
carpet the floor of this ancient
woodland in spring.

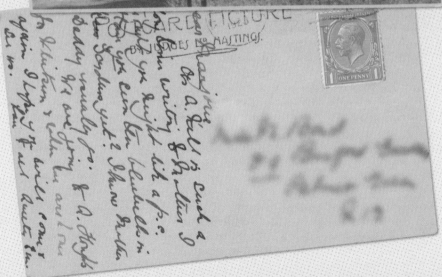

A rickety ladder is propped against the wall of the house by the station, a bucket of suds lifted aloft. A face appears at the upstairs window and the squeak of a chamois lets in the sunshine. A gardener with hose and mop washes the glass in Greenhouse Number Four. Yellow rays stream through the glazed roof, flowers turn their petals to the sun.

'Another lovely day. The window cleaners came this morning.'

This postcard was sent to St Leonards-on-Sea in September 1926. It shows Greenhouse Number Four, which was part of the T-Range group of glasshouses completed in 1869.

It is warm in the potting shed, the air smells of green things. He wipes his hands on his overalls, lifts the tray of sweet ladies off the bench and smiles back at them as they quiver in his hands.

'Weather good. We had a trip to Kew yesterday, trees and bushes gorgeous colours. Gardeners planting out beds of pansies.'

This postcard was sent in October 1964 and shows the Azalea Garden, which is planted in date order with hybrids from the 1820s to the present day.

AZALEA GARDENS, KEW

1284   KEW GARDENS.

Horseshoe beds are planted out with gaudy green spurge and sunshine yarrow. Pots of indigo verbena hitch a cart ride, waving as they turn past the pond.

This postcard shows the Pond at the end of the Broad Walk. It was written from an address in Kew Road and sent to the Hotel Placida in Montreux in the early 1900s. Montreux lies on the shore of Lake Geneva in Switzerland.

She would buy flowers,
if only she could –
a huge bouquet of
peonies wrapped in
crisp pink tissue.

THE HOUSE AND LAKE, KEW GARDENS.

**'Dear Annie I forgot that Tuesday
was your birthday – will have to
think of you another time when a
little better off.'**

This postcard bears the title: 'The
House and Lake', but in fact shows
the Pond and Museum Number
One, which was purpose-built as
a museum and has never been a
residential building.

The box is decorated with drawings of China – village scenes, mountains, tiered towers, golden dragons. She opens the bottle, breathing in heady oriental spices – aniseed, clove, peppercorns.

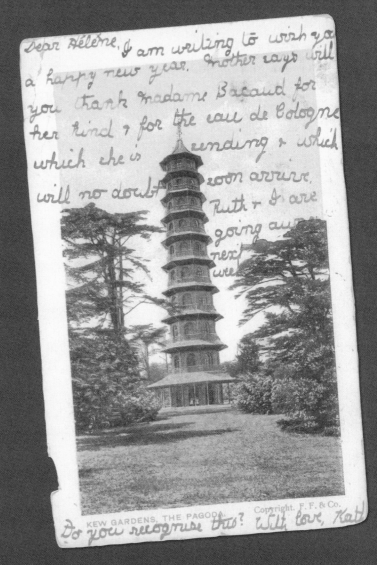

'Dear Helene, I am writing to wish you a happy new year. Mother says will you thank Madame Bacaud ... for the eau de Cologne which she is sending + which will no doubt soon arrive. Ruth + I are going away next week. Do you recognise this? With love, Kath'

This postcard showing the Pagoda was sent to the Boulevard de Clichy, Paris, on Christmas Day in 1905.

KEW GARDENS, THE PAGODA.   Copyright. F. F. & Co.

He pulls the golden ribbon
from the box, reciting the
poetry written on the little
card – *rose cream, strawberry
truffle, cherry delight.*

**'Thanks for the chocolate eaten
it have not died yet.'**

This postcard shows a view of
roses growing near the Palm
House. The Campanile can be
seen in the background. It was
built to disguise a chimney which
carried away smoke from coal
boilers in the Palm House and is
connected to the building via an
underground tunnel where a light
railway once operated.

THE ROSE PERGOLA NEAR THE ROCK GARDEN.
KEW GARDENS.

Gardeners inspect the roses straggling over the pergola, creaking with the weight of the musky blooms. Drowsy wasps crawl over ripening plums, sparrows cling with spindly toes to the spiked twigs of currant bushes.

The Rose Pergola was built in 1901 and runs through the Agius Evolution Garden and Kitchen Garden.

Rose petals blow across a tarmac path, pimpled by the heat. Planes bound for the airport fly in front of the sun – wheels out, noses down, thundering as they drop through the sky.

The Palm House, Kew Gardens

'It was quite exciting at the airport.'

This postcard showing the Palm House was sent in 1970, the year before UK decimalisation. The postmark is 'Richmond & Twickenham' and the stamp is a Queen Elizabeth II 4d stamp.

KEW GARDENS *from the Air*

As his magnolia petal parachute unfurls he sees his plane dive nose down, wings clipping elm trees. A burning, roaring dragon, scoring a scar in the turf on the vista.

Printed text on the reverse of this postcard sent in 1921 reads: 'An aeroplane view of Kew Gardens, showing the Pagoda, Temperate House and the Palm House.' There have been two plane crashes at Kew: in 1928 a single-seater Siskin aircraft crashed in flames near to Syon Vista; ten years later a plane towing an advertising banner made an emergency landing close to the Palm House.

V 4668   KEW GARDENS, THE LAKESIDE.

Springtime. An eastern breeze blows waterfowl across the Lake – Egyptian geese and their Canadian cousins, Bewick's swans, mallard. The sun looks at her winter coat and fur hat and shines more fiercely, determined to remove them. Greylags wade from the water, shake the feathers in their tails, stretch their necks to see what she has brought them. They peer into the crinkled paper bag as she unwraps the present she has made. A fruitcake diadem set with glacé jewels – green and red cherries, almonds, toasted seeds. Butter-brown and crumbling, just for them. In return she gathers their down as it tumbles over the grass, puffs it from her fingers and makes a wish.

This postcard shows the Lake. Throughout history, many birds have made their homes at Kew. Most famous of these is Joey, a Stanley crane who drowned in 1935 after falling through ice on the Lake.

45336   LONDON. SPRING IN KEW GARDENS

Clocks change, days spill over their edges, sun warms the soil. Spring shrugs off winter's woolly shroud and blooms. They peel off tweed and gabardine, telling each other, over and over, how warm it is. Everything is alive. New green shoots sprout from every cleft, pollen-yellow goslings follow their parents down to the pond.

This postcard sent in June 1950 shows the end of the Broad Walk. Kew reawakens in spring – flowerbeds are filled with daffodils and the trees are wreathed in blossom.

The Water Lily Pond, Kew Gardens.

Blackbirds and thrushes trill and flutter in the trees, plucking ripe berries from twigs, singing sweet songs for the lovers. Woodpigeons coo softly and fly down to the water's edge to drink, splashing the shallows over dusty wings. Bright green duckweed tresses tangle on the surface. In the shade of an old fir a pair of royal swans make their nest.

The Waterlily Pond on Cedar Vista was dug out in 1897 for tender aquatic plants such as waterlilies and irises.

THE ROCKERY, KEW GARDENS, LONDON. Copyright.

He waits across the street, under a plane tree, looking up at her window for hours. Smoking, fiddling with his pocket watch, trying to keep his nerve.

'I had meant coming to see you yesterday but on way out met a friend who begged of me to come to the music hall and I couldn't say no to such an offer.'

This postcard was sent in August 1907 to Bayswater and shows the Rock Garden, designed by William Thiselton-Dyer, Kew's third director (and son-in-law of Joseph Hooker, Kew's second director). The Rock Garden was built in 1882 to showcase plants from mountainous regions.

He couldn't leave without her. They made a new life in the cool Malaysian mountains, shrouded in mist and green camellia. Every Saturday night his new wife clips her hair into a delicate wooden clasp shaped like an orchid. She smiles at him as they sway across the dancehall floor to records scratching through a golden gramophone trumpet.

'Lady Mullaly is going to ask Hugh's brother, Dick, to look you up. He is on leave – a Malay Tea Planter – A.1. dancer.'

This postcard shows the Duke's Garden behind Cambridge Cottage, the former residence of the Duke of Cambridge. Large plantations of tea grow in the highlands in north-west Malaysia.

DUKE'S GARDEN, KEW GARDENS

In the function room of a riverside pub, Mr Val Bayles counts *five, six, seven, eight* and spins Lollie across the beer-stained floral carpet. He tells her, breathlessly, how he had been the darling of London music halls; how on the boards of Paris stages he had been adored.

Kew Gardens, The Palm House.

'I expect you will be surprised to hear that I am going to Mr Val Bayles' dancing classes. I have talked of it many times, but I am going tonight for the first time. When you have a dance in the shop I shall not feel such a duffer … With love from your affectionate friend Lollie.'

This card was sent from Twickenham to Isleworth in November 1909 and shows the Palm House.

There's treasure inside. Baskets and string bags woven from rushes, wooden bowls, flasks made from gourds, necklaces of enormous seeds strung on plaited straw. Barkcloth shirts, paddles for boats, a bamboo flute that sounds like a bird. Arrows and bows. Boxes carved from tropical wood – tiny chests filled with nutmeg and mace. Glass bottles of pickled roots, jewelled leaves drowned in vinegar and alcohol.

Before being converted into the School of Horticulture, the building shown on this postcard was Kew's Museum Number Two or Reference Museum. The Aquatic Garden in the foreground was built in 1873 and replaced in 1909 by seven new tanks.

The Herbarium – an old building with an old name. A plant library, poked into a corner of the garden – a paper ticket forgotten in a pocket. The place smells of history – must and polish. Specimens are stuffed stiff into cabinets, parched and brittle. Hidebound ledgers piled on mahogany desks hold thousands of alphabetically ordered entries. Vast plan chests are filled with charts of the oceans and maps of the land, where men wearing top hats and mutton chops plundered dry savannah and freshwater swamp, hunting plants until they were caught and caged. Not with spears and stealth, but with scissors, sketchbooks and blotting paper, with cases and campfires, with presses, vasculums and microscopes, with glue made from rabbit guts.

Kew's Herbarium contains seven million specimens of dried plants from all over the world, including some from the collection of Charles Darwin. The building is a Grade II listed structure. The floors are galleried with a void in the centre to let in as much natural light as possible because candles – commonly used elsewhere before the advent of electric light – were not permitted in the building as the fire risk to the dried collections was too great.

ROYAL BOTANIC GARDENS, KEW
Spring Scene; with Japanese Cherry (*Prunus serrulata*) and varieties of *Narcissus* (Daffodils, etc.).
Copyright, J. Arthur Dixon Ltd.   Printed in Great Britain.

The botanists wear broad-brimmed hats and sail the oceans, seeking to order the world and make sense of it, to conquer knowledge – to harpoon it, catch it in a net. They collect seed-pods from a coastal forest and pick fibrous stalks to weave into new mast rope. They snip and scythe, pulling clinging plants from the trunks of trees in dense forest, cutting specimens from a soily slope, pickling them in island rum.

'Writing this, sitting on a Kew bench. The sun has been lovely all day while we have been strolling around the Gardens … We also saw an exciting film on at the cinema. It was about a ship wreck. Mary and John .'

Printed text on the reverse of this postcard sent in 1973 reads: 'Spring Scene; with Japanese Cherry (*Prunus serrulata*) and varieties of *Narcissus* (Daffodils, etc.).'

Kew's iconic Palm House was built in 1844 by Richard Turner to Decimus Burton's designs, which borrowed from ship-building techniques – hence the building resembling the upturned hull of a ship.

13097

The Great Palm House, Kew Gardens.

The boy points across the lawn, insisting he can see a ship. A huge glazed galleon carrying a jungle cargo, sailing through the garden on a sky-blue ocean. His father takes him by the hand and together they climb aboard, staring up into the metal cage of ribs and rivet heads – hammer forged, welded, polished, breathing. Joists and cantilevers fill the upturned hull. Girders, grilles, gratings painted with torpedo varnish wear two coats, as recommended by the Corrosion Advice Bureau. Wrought, cast and soldered, fused together with knowledge and pluck. A mighty ark, afloat.

He stands barefoot on the shore. Waves run up the beach, planting her sopping kisses on the hem of his trousers. He writes her name in the sand and watches a crab dance sideways into the water.

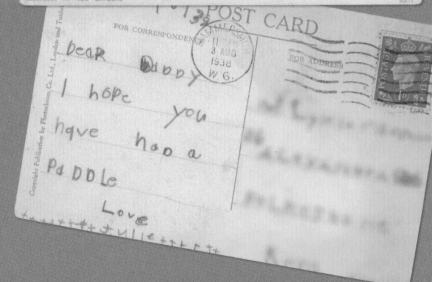

'Dear Daddy, I hope you have had a paddle. Love xxxxxxxx Julie xxxxxx'

This postcard showing bluebells growing in the Natural Area was written by a child and sent in August 1938 from Hammersmith to Folkstone with a George VI 1d stamp.

THE BROAD WALK, KEW GARDENS

Benches line the Broad Walk. Each tells a story, embossed on a metal plaque. *For our colleague, who gave advice but seldom took it. To remember Mum and Dad who loved these gardens, and each other. My husband liked to sit here – I miss him.*

This postcard shows the Broad Walk, a wide avenue connecting the Orangery and the Pond, designed by landscape architect William Nesfield and originally laid out in 1845.

THE MUSEUM AND LAKE, KEW GARDENS

People loll across the seats, delving into bags for confections and
juice. A bin brims with the day's detritus. A squirrel clambers in
to feast on discarded things: crusts and apple cores, carrot batons,
scraps of chicken, salt left in crisp packet corners. Quarrelling
starlings gather as boys push chubby fingers inside humid
polythene bags, feeding the fish stale seeded wholemeal.

Two postcards showing
flowers growing in the Palm
House Parterre, the Pond
and Museum Number One.

She adjusts her scarf, watching the *Pick Your Own* sign recede in her oblong mirror. The smell of tomato vine mixes with dusty road and her round red passengers bounce on the back seat.

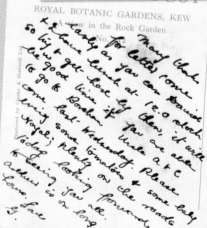

'Come as early as you can. David and I have lunch at 12 o'clock so try to get here by then ... Please bring some tomatoes and some baby royal; plenty on the roads today.'

Printed text on the reverse of this postcard sent to a Suffolk village reads: 'A View in the Rock Garden.'

They tie hairy string to the bottles and slip them into the cool water. It gulps them down. Beer froths inside the heavy brown glass, water boatmen scull away into the rushes. She spreads a tartan car rug in the long grass under a pine tree, feels for hard boiled eggs in the darkness of the hamper and passes them to him. He peels away the sharp shards of shell. Jays chase through the conifers, laughing at the meaty green finger cones that ooze thick resin like fat from cold sausages.

'We are sitting just about the pine on the other side. We are just about grilled.'

This postcard was sent in June 1960 and shows the Lake, which covers five acres and has four islands. It was created in 1856 and the gravel excavated was used for terracing at the Temperate House.

The dry earth heaves and shifts. Fractures appear along the edges of the Broad Walk, where the earth is parched, the grass bleached to flaxen straw. The sun has scorched the ground with its fiery temper, baking a cracked crust on the shortbread lawns.

'We have struck a heatwave.'

Printed text on the reverse of this postcard sent in 1969 reads: 'View along the Broad Walk, with beds of tulips and rhododendrons ('Lord Palmerston', deep rose; 'Hugo de Vries', pink); trees of the yellow-leaved variety of the 'Red Oak' (*Quercus rubra*), the 'Red Horse-Chestnut' (*Aesculus carnea*) etc., in the background; in the distance, Kew Palace, and nearby a fine Oriental Plane (*Platanus orientalis*), planted in 1762.'

Kew Palace and Sun Dial, Kew Gardens.

The sun burns so fiercely on the sundial it erases itself, throwing no shadows at all, singeing jagged holes in history's papery pages. Time melts, becomes translucent. It shifts and wobbles like the space above a candle – a wavering mirage. Hours expand and float away. Light is a line to travel along – a luminous elastic tightrope stretching back into the past.

The sundial on Kew Palace lawn sits on the site of a former royal residence known as the White House, which was demolished in 1802.

Plants as old as the Pantheon grow around the temple. A child piles sticks on the steps – offerings at a shrine.

King William's Temple was built in 1837 for Queen Victoria, in memory of William IV. It stands on a mound in the Mediterranean Garden and is surrounded by olive trees, lavender and cypresses.

Eleven o'clock. A fair-haired boy escapes the clutch of his mother. He sees the hill, wants to climb it – to plant his flag and shout from the top, *I'm here, look at me!* A pathway winds upwards through nettles and wild garlic. Reaching the summit, gasping, he looks down on his kingdom. *I am the god of the winds! I am the king of storms! Look at me!* He puffs his cheeks, blowing with all the breath in him, whistling it down the hill into an animal hole.

The Temple of Aeolus was built in 1845 to Decimus Burton's design, based on a previous version built in the 1760s to Aeolus, the ancient Greek god of the winds. The original temple had a revolving wooden seat that afforded a circular view of the surrounding landscape.

While the others pose for photographs, Sonny plunders the garden for treasure: a swan-white feather, an oak leaf, a woodlouse rolled into a pill ball, a piece of lichened bark.

'Good morning Lolo! Are you glad to be home again? Lots of kisses from Sonny.'

This postcard was sent to Passy, a district of Paris. It shows the Temple of the Sun, which was destroyed in March 1916 by a cedar tree which fell during a storm.

POST CARD.
(For Address Only.)

good morning
Lolo! are you
glad to be home
again. lots
of k

Photo Cole & Polden
Temple of the Sun, Kew Gardens.
13100

Kew Gardens. Museum Nº 3. — LL.

The branch of the cedar creaks under the weight of the man with the camera who looks through his viewfinder and finds them turned upside down.

This postcard shows the Orangery, which was converted into a timber museum in 1863. The initials 'L.L.' refer to the French photography and printing company Léon & Lévy, which specialised in picture postcards.

22266 Kew Gardens. The Entrance.

The rules are stuck to the gate: no swimming in the ornamental water or defacing the statues. No swearing, no frightening the ducks. No eating the fruit growing in the glasshouses or plucking petals from the flowers. No climbing trees – even when no one is looking.

This postcard sent in May 1908 shows Elizabeth Gate. Kew's Statutory Garden Regulations can be enforced by the Kew Constables who have police powers within the Gardens.

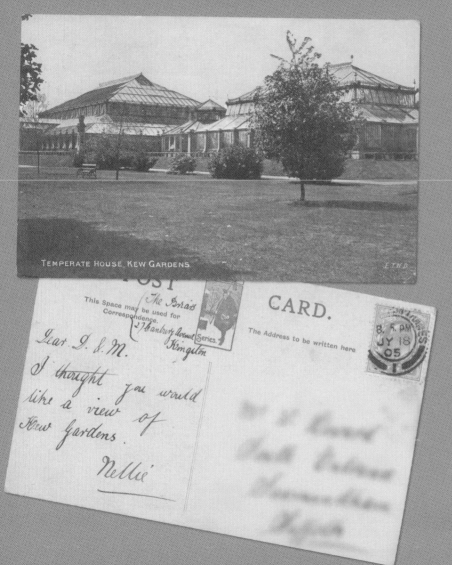

TEMPERATE HOUSE, KEW GARDENS.

E.T.W.D.

POST CARD.

This Space may be used for Correspondence.

The Briars
27 Banbury Avenue
Kingston

Series. 1

The Address to be written here

8.5 AM
JY 18
05

Dear I. & M.
I thought you would
like a view of
Kew Gardens.

Nellie

The colossal columns of the great glass cathedral are louder than an organ. Plants fill the nave, spilling umbrella fronds onto grilled floor panels hiding pipes leading from the undercroft. Once, day and night, men worked like ruddy faced devils, feeding the roaring crypt furnace to keep the plants warm.

'I thought you would like a view of Kew Gardens.'

This postcard was sent in July 1905 and shows the Temperate House, the largest Victorian glasshouse in the world, designed by Decimus Burton. Construction began in 1859 and took 40 years to complete. Terracotta urns on the outside of the building conceal chimneys previously used to release steam from the underfloor heating system.

They give their coins to the man at the gate and stand with their backs to the bonfire until they are too warm and wonder aloud if their hair might set alight. A rocket screams, showering sparks and reflecting the Pagoda in their eyes. The air smells of spun sugar, chestnuts and cinnamon.

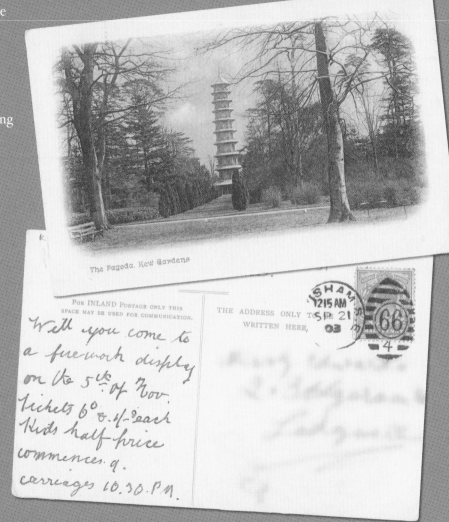

The Pagoda. Kew Gardens

'Will you come to a firework display on the 5th of Nov? Tickets sixpence and one shilling each. Kids half price.'

This postcard was sent in September 1903 from Lewisham to Ladywell and shows a view of the Pagoda, designed by William Chambers to emulate a Chinese pagoda. It was completed in 1762 and originally decorated with 80 brightly painted and gilded dragons which have now been restored.

230   LONDON W. — Kew Gardens. — The Tea House. — LL.

The women sipped Earl Grey from dainty china cups, whispering about fairness and flames. How easy it would be to set this place alight. The brigade came quickly with their helmets and hoses, pouring water which hissed and spat on the burning wood. Charred oak creaks in the wind and the women weep in their cells.

In February 1913 suffragettes smashed the windows of three of the orchid houses and damaged plants. They left behind an envelope bearing the words 'votes for women'. Twelve days later the Pavilion was burnt to the ground in an arson attack and two suffragettes, Olive Wharry and Lilian Lenton, were arrested and imprisoned.

Like grimacing, grunting wood pigs, tusks braced over top lips, the men grub out hellebores with their bare hands, heaving the matted hairy bauble from the soil – lifting, tugging, wrestling with the tree until they win. The horses watch silently, breathing hot steam into frosted air.

Printed text on the reverse of this postcard reads: 'Barron Tree Transporter near the Palm House at the end of the 19th Century. The Transporter may now be seen by the Tea Bar.' William Barron's Tree Transporter was purchased by Kew in 1866 and used by Lancelot 'Capability' Brown to remodel landscapes by uprooting mature trees and moving them to different locations. It was nicknamed 'The Devil' as it injured several gardeners.

The sandy pathway is rutted where rims have cut grooves over centuries. A crested peacock feels the rumble of wheels and sweeps away through the gate, screeching an insult over the wall.

Printed text on the reverse of this postcard reads: 'Tulips outside Cambridge Cottage Garden.' Cambridge Cottage Garden is now referred to as Duke's Garden and the flowerbed shown in the picture is known as the Duchess Border, to commemorate Augusta, wife of the first Duke of Cambridge. Horses (mostly Shires and Suffolk Punches) were used at Kew until 1961.

In the evening the men return home from the forest smelling of clean sweat. They rasp hands across the shadows on their faces and produce from their pockets comfits so sour they give their children dimples. They tell bedtime stories – how they conquered the giant by sawing at his ankles until he bled hot resin, how they shouted when he fell. How they dragged him through the canyon and sent him away down the logging road.

Printed text on the reverse of this postcard reads: 'The Flagstaff. A spar of Douglas Fir 214 ft high, from Vancouver Island. Presented by the Government of British Columbia in 1919.'

Here is the tree, by the path near the gate. The metal label clinks in the wind, singing its name: Corsican pine. *So far away from home*, she says, as she stretches out her fingers to stroke the bark before slowly turning to leave.

'Had an enjoyable day at Kew, Hampton Court, Richmond and then sailed down the river to Westminster Bridge. I shall be sorry to leave for home on Monday as I have had a very pleasant time.'

Printed text on the reverse of this postcard sent in October 1925 reads: 'Corsican Pine (*Pinus laricio*). Brought to Kew from the south of France in 1814. Probably the oldest tree of its kind in Britain.'

Back from her walk through shaded lanes, Elsa steps out onto the balcony of her room at the Hotel Majestic. She watches riders canter their horses round the sweep of the bay. Dinghies flutter coloured sails, paddling children dig holes in the sand for razor clams. A breeze blows – salt from the sea, the scent of hot pastry, dried magenta petals of bougainvillaea.

KEW GARDENS
The JAPANESE GATEWAY

'My dear little Lucia, I was indeed surprised to get your letter after all that long time! But on the other hand 'better late than never' and I am pleased to see you have not quite forgotten me. I am glad you are having such a lovely time and that you are all well. We are going away on the 1st of July for 4 weeks – first to Spain and then to La Baule to meet Auntie Anna, Andre and Elsa. When are you coming back? Heaps of love and kisses to all from us 3 to you. From Auntie Jojo'

This postcard showing the Japanese Gateway was sent to Vienna, with a George V three halfpence stamp in June 1921. La Baule-Escoublac, known as 'La Baule' is a luxury seaside resort on the Côte d'Amour in north-west France.

The Pagoda from the Palmhouse, Kew Gardens.

I hope that you will visit this place, where tiny seeds germinate, filling flowerbeds with a thousand pink petals. Bushes grow like macaroons; wisteria confetti blows across the surface of the Lake into the locks of the maidenhair tree. Pine cones swell duck egg green, bees buzz in turquoise hives. Bamboo grows thick in the hollow and lilacs sway at the edge of a lawn.

'J'espère que bientôt vous visiterez cette place. Quand je parte pour Allemande je vous ecrirerai.'

['I hope that you will soon visit this place. When I leave for Germany I will write to you.']

This postcard sent in February 1946 to Brussels and entitled 'The Pagoda from The Palm House' shows the landscape at the back of the Palm House, and a view down Pagoda Vista. It is written in poor French. A better construction would be: *J'espère que vous aurez l'occasion de visiter cet endroit. Je vous ecrirerai lors de mon depart en Allemagne.*

The Rhododendron Dell, Kew Gardens.

POST CARD

(FOR ADDRESS ONLY.)

Dear Myrtle:-
Well everything
I am in London,
getting along n
this morning I
for a passport
N.Y. Give m
to everyone and I'm d
sorry I was unable to come
and see you all Saturday. In
a day or two when I am mon

1361

Printed in England by GALE & POLDEN, LTD.
London, Aldershot and Portsmouth. (Copyright.)

She prints her name on the dotted line and hands the form to the desk clerk who embosses it loudly with his stamp. The sound echoes round the marbled walls and falls out the window, startling the pigeons roosting on the ledge. They fly away across the rooftops, into the garden, stealing through the air and over the border, free, unfettered.

'Mother is getting along nicely now and this morning I made application for a passport to return to N.Y. ... In a day or two when I am more settled I'll write and tell you about everything. Nellie.'

This postcard showing a painting of the Rhododendron Dell was sent to New York in May 1920, to an address on Lexington Avenue, which stretches for five and a half miles down the east side of Manhattan.

Beneath the gabled terrace, black canal water slaps the side of the holiday houseboat, jolting the saggy bed. Barbara groans and rolls over, lamenting the size of her dinner. The pair of them – two steaming bitterballen – had rolled from bar to bar, bouncing off brown wood panelling. They held herring by the tails high above their mouths, snatching them like harbour seals coming up for air. They ate poffertjes, inhaling dusty sugar and gobbled raw ox sausage spiced with pepper and nutmeg. Rubber yellow gouda triangles were washed down with gin in tulip goblets that reminded them of the flower fields out on the eastern marshes.

'Had a day out at Kew. Beautiful flowers and weather. Have recently had holiday in Belgium and Holland with my sister Barbara. The food was gorgeous – put on a stone. The Tulip fields were a wonderful sight.'

This postcard shows a view of the Palm House and Parterre. The Queen's Beasts can be seen in front of the building. These ten heraldic figures were installed in 1956 and include the White Greyhound of Richmond, the Black Bull of Clarence, the Falcon of the Plantagenets and the White Horse of Hanover.

In the evening, Lady Bunbury asks for her bath to be drawn and filled with herbs – sweet tansy, feverfew. She unbuttons her brass fastenings, slinks out of her shot silk gown and steps, pink and delicate, into the hot water. Beatie kneels behind, pouring a jug to rinse the soap suds.

THE MUSEUM, KEW GARDENS.

'c/o Lady Bunbury … Dear Sister I am writing a P.C. to tell you I have got a situation in Suffolk you will see by the address. With my very best love, hoping to hear from you soon. Good bye dear, Beatie.'

This postcard was sent in July 1909 and shows the Orangery, then Kew's Timber Museum. A large collection of timber and furniture woods from the Great Exhibition were transferred to the Museum in 1863, the year after it opened. Galleries and spiral staircases were added to the inside of the building in 1883.

Kew Gardens *Lundi*

a large green house where it is hot!!! Kew!!!

Much love *Hélène*

She ascends the wrought iron staircase, treading lightly on each
step, thinking of all the people who have climbed up before her,
all the laced boots and hitched skirts. It is humid and her hair
hangs lank around her face. On the balcony, she stares down at
the plants, realising how small she is, and how young.

'A large green house where it is hot!!!'

This early postcard shows
the Palm House and Pond.
It was written on a Monday
('Lundi') by Hélène.

The smell inside the Waterlily House is intoxicating
– sweet pineapple. The flowers unfurl one by one
in the wet air – unzipping green leather jackets,
blooming moon huge.

Giant amazon waterlilies flower at night and
only last for 48 hours. The petals are white
on first opening, turning pink on the second
night. In the wild they are pollinated by
beetles but at Kew they are hand pollinated
by staff working special night shifts.

Grandpa pulls a tattered paperback from his jacket pocket and puts his nose in it, pretending not to watch the people around him gagging themselves with sticky buns and spilling tea in their saucers.

'Just returned from the gardens, our usual constitutional – Grandpa feeling better today.'

This postcard shows the Pavilion, described as 'The Tea Rooms'. It was sent to Seasalter.

POST CARD.

THE ADDRESS ONLY TO BE
WRITTEN HERE

Kew Gardens.—The Lake.

60

Two leaves from the maidenhair tree. One from each of them, peeled from the damp pathway after the rain. Grandma finds her address book in the pocket of her snap shut handbag and presses the golden fans carefully between the pages, under *G* for ginkgo. Trapped, like fossils in shale between Dr Gould and Cecily Green.

'We went into this park on Friday afternoon with Grandma. It is a lovely place. How you would like it. We have two leaves for you from the Maidenhair tree.'

The *Ginkgo biloba* tree was planted at Kew in 1762, near to the Orangery. This postcard was sent in 1907 when the tree was 145 years old.

Variegated ivy and
marbled creepers
cling to crumbled
brickwork, strangling
the ruin.

'Am just half way through my
exam. It has not been bad so
far. Apart from the anxiety I
enjoy the life.'

This postcard was sent to a
village in Cambridgeshire.
It shows the Ruined Arch,
built in 1759 as a folly that
also served as a bridge over
which sheep and cattle were
brought from Kew Road into
the Gardens.

THE RUINED ARCH. KEW GARDENS.

Marianne North Gallery, Kew Gardens

A schoolgirl sketches with a pencil, copying a painting on the wall made by a woman who cut her way through wet vine curtains with her bone-handled knife, fording rapid rivers and sleeping in a hammock, listening to nightjars and crickets clicking as she fell asleep. A woman who painted trees and mountains and a tiny green songbird – casting a spell of shape and colour with her brush.

This postcard shows the Marianne North Gallery. Marianne North travelled the world, painting pictures of plants and local habitats wherever she went. In 1879 she persuaded her friend, Kew director Joseph Hooker, to build a gallery to display her work.

Enid concentrates hard, pushing her glasses up her nose, staring at her teacher's chalky drawing, imagining the faraway places. Hawaii, the Himalaya, the coast of New South Wales. A tiny island in the South Atlantic Ocean, the mountains of Taiwan. Remote wildernesses where the animals have pockets and geckos swallow black butterflies, where forest birds sound like an orchestra tuning up. Places where mountain parrots swoop through dense valleys and hanging gardens of epiphytes thrive in the pockets of trees. Where plants grow on shingle in the channels of a braided river on the island of a thousand tongues.

'Dear Enid, I am sending you a No. 6 as promised. I hope you were a good girl in school yesterday.'

This postcard was sent from Clapham in 1905 and shows the interior of the Temperate House, which contains some of the world's rarest and most threatened temperate plants from five different continents.

Interior of Winter Garden, Kew Gardens.

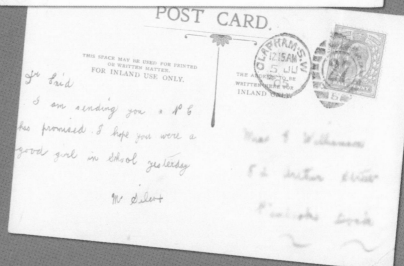

LONDON. THE LAKE, KEW GARDENS.

Dea. J.

Will you ask Bert to get another album for p.p.c's like the last, red if possible. Will it be convenient for Plowright & I to run over & fetch it on Saturday night. It is for him. Hope you are both well & your cold gone. Saw Aunt last night & the Dorah. She was very gay. Has a slight cold. Yours BB.

Bert sits in his wing chair waiting for the postman, listening for footsteps on the path – the squeak of the letterbox, the smack of cards on the bristly coir mat. He sticks them into his album, steadying his hand as he admires the views of Cairo and Provence, where oranges and lavender grow – places he will never go.

'Will you ask Bert to get another album for p.p.c's like he had, red if possible.'

This postcard shows Museum Number One and the Pond (described as 'The Lake'). It was sent to Clapham Common in March 1903. 'p.p.c's' are picture post cards.

KEW GARDENS, JAPANESE GATEWAY.

V4667

She sits with her sister, close enough to feel the warmth of her body which she has known since before they were born. Identical, save for their tempers and their hair – chestnut and ash. She plucks yolk yellow buttercups and stares at the gateway, imagining a land where sandalwood smoke drifts through the paper doors of a temple, where cherries and maple grow on the banks of the shallow river and a geisha-pale crane creeps on black lacquered legs.

Chokushi-Mon, the Gateway of the Imperial Messenger, or the Japanese Gateway is styled on the Gate of Nishi Hongan-ji temple in Kyoto. It was originally created for the Japan–British Exhibition held in London in 1910 and transferred to Kew the following year. It stands on Mossy Hill, the site of an earlier mosque replica (hence Mossy).

50213. KEW GARDENS. RHODADENDRON DELL.

The soldiers labour in their regiment red, digging out the hollow – cutting through layers of earth, pulling the worms up to look at the sky, changing the land, sculpting history.

The Rhododendron Dell was designed as Hollow Walk in 1773 by Lancelot 'Capability' Brown and dug out by soldiers of the Staffordshire Militia from land that was previously incredibly flat.

I. KEW PALACE. EXTERIOR.

H. M. Office of Works.

On his allotment, Arthur plants vegetables in neat rows, growing food for his country. A squadron of green onions turn their hairy heads to salute him.

'We took the opportunity whilst on holiday last week to visit Kew Gardens. I hated returning to work on Monday. We are having very disturbed nights but fortunately nothing has dropped near us so far. Glad to hear you are O.K. + I hope to see you when you come to London. I shall be on L.D.V. duty all day on Sunday 8th Sept. I am also on tomorrow. All the best. Arthur'

His Majesty's Office of Works (which printed this postcard sent in August 1941) was established in 1378 to oversee building maintenance of royal residences such as Kew Palace, shown. The L.D.V. was the Local Defence Volunteers, an early name for the citizen army that became the Home Guard in World War II. Bombs fell in several areas of the Gardens during that war, including on Syon, Pagoda and Cedar Vistas, and on Boathouse Walk and Holly Walk.

THE ROCK GARDEN, KEW GARDENS.

Printed in England by Gale & Polden, Ltd
London, Aldershot, and Portsmouth

Edie fetches the brass polish from the understairs cupboard to shine the fleur-de-lys on her uniform buttons. She hums a tune from the musical, wishing she could wear the costumes instead.

'Bessie is quite right when she says we are not free. In England the WAACs cannot wear hat pins, army stockings, not light ones … Mustn't wear army kind of shoes, must wear proper collars, it's nothing but rules here. Last Saturday two of us went to see Oscar Asche in Chu Chin Chow, it was very good, the scenery and dresses were gorgeous. Sunday we went to Kew Gardens. It stretches over a great piece of ground, beautiful trees, and hot houses where they have palms and trees of every country. Edie'

This postcard showing the Rock Garden was written in April 1919, after the end of World War I. WAAC was the Women's Auxiliary Army Corps, the women's division of the British Army during the war and in the years immediately afterwards. One of the rank insignia worn on uniform epaulettes was the fleur-de-lys. *Chu Chin Chow* was a musical comedy by Australian actor, writer and director Oscar Asche, based on the story of *Ali Baba and the Forty Thieves*. It showed at His Majesty's Theatre in London for five years between 1916 and 1921.

He sits up high, in a plush velvet seat near the grand chandelier in the theatre's ceiling, watching the glimmer of carefully cut crystals. A stagehand pulls a rope to move the scenery and a painted garden descends from the roof.

'Maurice dear, I was so pleased to receive your letter, I'm wanting an opportunity to write a long letter telling you all about Covent Garden, my trip to Jersey by plane and the many beauty spots in and around London I've seen lately. Last Saturday we went to Kew and the bluebells in the woods were just glorious. It was so lovely to be amongst them and listen to the song of the birds. Last Sat evening went to an operetta by Noel Coward. The cast included Irene Vanbrugh. Will send the programme later. I was disappointed indeed to find an artist of her quality in this type of show.'

This postcard shows the Duke's Garden, described as 'Cambridge Cottage Garden'. It was written on 25 May 1938 and sent to Australia, where it was received on 27 June of the same year. *Operette* was a Noel Coward musical that showed at His Majesty's Theatre in London in 1938 and starred English actress Dame Irene Vanbrugh DBE.

A mantel clock ticks time in the drawing room,
where oil portraits of botanical luminaries keep
watch over vases which echo with dried hydrangea.
Dust motes fall on the piano. Beyond the stiff
brocade curtains, water slides from a nymph,
tickling fish in the pool below.

The Duke of Cambridge (grandson
of George III) lived in Cambridge
Cottage in the late 1800s. After
his death in 1904 the cottage was
converted into the Museum of British
Forestry (Museum Number Four)
which opened in 1910 and in 1958
became the Wood Museum.

Kew Gardens, The Rockery.

The hours between sunset and sunrise are stretched taut, pulsing on a buckskin drum. The garden lies still, holding its breath. The moon looks in through the open doors of sheds, reaching to touch the blades of scythes. Animals come out to play. A badger rustles the bushes, the darkness echoes with the shriek of a vixen, betraying her hunting place in the rockery where she stalks the smell of rats – creeping on her haunches, tumbling squeaking prey until fear clots in tiny veins.

'H. Rayner here to discuss last night. Came over in his motor.'

This postcard showing the Rock Garden was sent from Sydenham to Bristol in 1904. It would have offered a rare opportunity to see what the Gardens looked like at night.

Lion Gate. Kew Gardens.

Bored, they creep up on the lion, pretending to stalk him. They skulk outside the lodge. Chimp fingers grip the railings, noses push through the bars, as they wonder who lives there. Who lights the fire in the fireplace? Who sleeps in the saggy iron bed? Who tiptoes out the door in their nightshirt to wander through landscaped labyrinths where topiary figures grin at the moon?

This postcard shows Lion Gate Lodge, built in 1863 on the same site as another lodge constructed just 12 years earlier for the foreman of the 'Pleasure Grounds'. The gate was originally Pagoda Gate and the Pagoda can be seen in the background. The Lion (together with the Unicorn that now guards Unicorn Gate on Kew Road) originally sat atop the entrance to an old gate on Kew Green.

The Rhododendron Dell, Kew Gardens.

13108

POST CARD

(For Address only)

Printed in England

She licks the back of the king's head and sticks him in the corner. King George of the United Kingdom and British Dominions. His Imperial Majesty, Emperor of India, ruler of elephants, crocodiles and handsome spotted leopards. At the zoo, a Royal Bengal tiger lies against the wall of his house, flicking his tail to shoo away the flies. He closes his eyes, remembering the forest where he padded softly through thickets of rhododendron.

'I am having a fine time here. I was round this way on Sunday afternoon after ... Buckingham Palace. This is a grand place ... Spent hours at the zoo yesterday.'

This postcard sent to Fife shows the Rhododendron Dell. Kew's second director, Sir Joseph Hooker, brought back rhododendrons from his travels to the Himalaya in the mid-1800s.

BLUEBELLS.

ROYAL BOTANIC GARDENS, KEW.

45:

At the bottom of the garden, the air smells of deep woods and the memory of something darker – meat and musk. Down in the bluebells, the king hid his zoo. A wild menagerie, stuffed with animals plucked from their burrows, stunned with darts, lured into traps. Kangaroos with babies in their pockets, exotic pheasants, springbok, a brace of black swan, a giant tortoise with a cracked shell.

Kings George II and III both kept a menagerie at Kew, in what is now the Natural Area, where bluebells grow. Exotic animals fetched from abroad were symbols of wealth and status and were popular among royalty and the aristocracy.

KEW PALACE.

Here is the palace – a doll's house built for a king. The doctor locks him up faster than the sugar and spice in the kitchen clerk's office and bleeds him with leeches winkled out of the lake. His trapped mind flutters against the glass in the dining window panes like a pigeon caught behind the wainscot. His fingers pick at the textured curtains as he looks out towards the river, weeping and wigless, croaking music he learnt as a boy.

Kew Palace was used by King George III as a countryside retreat. He suffered from mental illness which royal physicians treated with inhumane methods. This early postcard also shows the Orangery.

At the edge of the garden, rolling lawns studded with stars give way to the sweep of river bend which holds them in a crooked embrace. Fluently, grass becomes a shingle shore – the boundary is blurred, land becomes liquid.

Printed text on the reverse of this postcard reads: 'China Asters near the River.' Syon Park and House can be seen in the background. Lancelot 'Capability' Brown designed the park in 1760 and was subsequently commissioned by George III to landscape parts of Kew. He installed the ha-ha (a ditch in front of a wall) that borders Kew on the river side.

"A SOWER", KEW GARDENS.

Pampas grows tall and feathery, dusting off clouds above the Grass Garden. The sower throws his seed into the breeze blowing off the river.

'Enjoying our holiday very much. The weather has been good not too warm. Was here today it's a beautiful place. Sailed back to London on the Thames, it was grand. See you soon. Love to all.'

The sculpture 'A Sower' by Sir Hamo Thornycroft can be found in the Grass Garden. It was cast in 1886 and was a gift from the Royal Academy. The pedestal was designed by Sir Edwin Lutyens.

Crowned copper faces
shimmer up through black
water, winking from the
bottom where people have
thrown them, for luck.

'Kew out from London 28th July 08. Dear George & Mary, I am getting splendid weather. Father.'

This postcard was sent in July 1908 and bears a Kew Gardens postmark. The Waterlily House was built in 1852 specifically to showcase exotic waterlilies. It is the hottest and most humid environment at Kew.

KEW GARDENS, THE ROCKERY.

Copyright. F. F. & Co.

She presses the button labelled 5, feeling the excitement in her throat as she rides the lift up to ladieswear. In the changing room she gazes at her reflection – a pretty fritillary, mottled and milky, in a skirt dyed gentian blue.

'Dear E, I will be over this evening, about ½ past 6 to fetch your coat. I thought I had better let you know, in case you were out shopping. With love from B. Excuse scribble in haste also bad pen.'

This postcard showing the Rock Garden was sent from Leytonstone to Snaresbrook in May 1909. Instructions on the reverse of the card read: 'For inland postage this space, as well as the back, may now be used for communication. For foreign postage the back only (Post Office Regulations).'

Valentine's Series.   The Museum, Kew Gardens.   *July 2/4 G.H.G.*

The green line of the District Railway ribbons across the city – a tendril curling around the edges of allotments, a root crackling through dark tunnels. Turnham Green to West Kensington. He sinks into the leather seat, inventorying the stations in his head, counting them off as he stares at the tattered poster which says *Kew Gardens – Go By Underground*.

'Am going to help them take stock at West Ken tomorrow. They start at 4am but I don't think I shall get there quite so early. Hope you are all well.'

This postcard bearing a Kew Gardens postmark was sent in July 1904 and shows Museum Number One, described as 'The Museum'. Kew's second director, Sir Joseph Hooker, had wanted Kew Gardens station to be built opposite what is now Temperate House Gate and lowered the wall in preparation, however the station was eventually built half a kilometre further away.

Synth pop falls in spangled neon shapes from the conservatory's mess room window. Gardeners on their lunch-breaks shimmy across the tiled floor of the disco kitchen, adding a microwave ping to the beat. An opera of pitcher plants sings flat and loud through rubbery red mouths and desert cacti dance with fists raised, punching holes in the zigzag roof.

These postcards from the 1990s show views of the Princess of Wales Conservatory, opened in 1987 by Diana, Princess of Wales, and named for her predecessor, Augusta (mother of George III) who enlarged and improved the Gardens in the 1700s. Printed text on the reverse of the postcard showing the interior reads: 'The dry tropical zone. One of ten computer-controlled climatic zones contained within this striking example of modern glasshouse technology.'

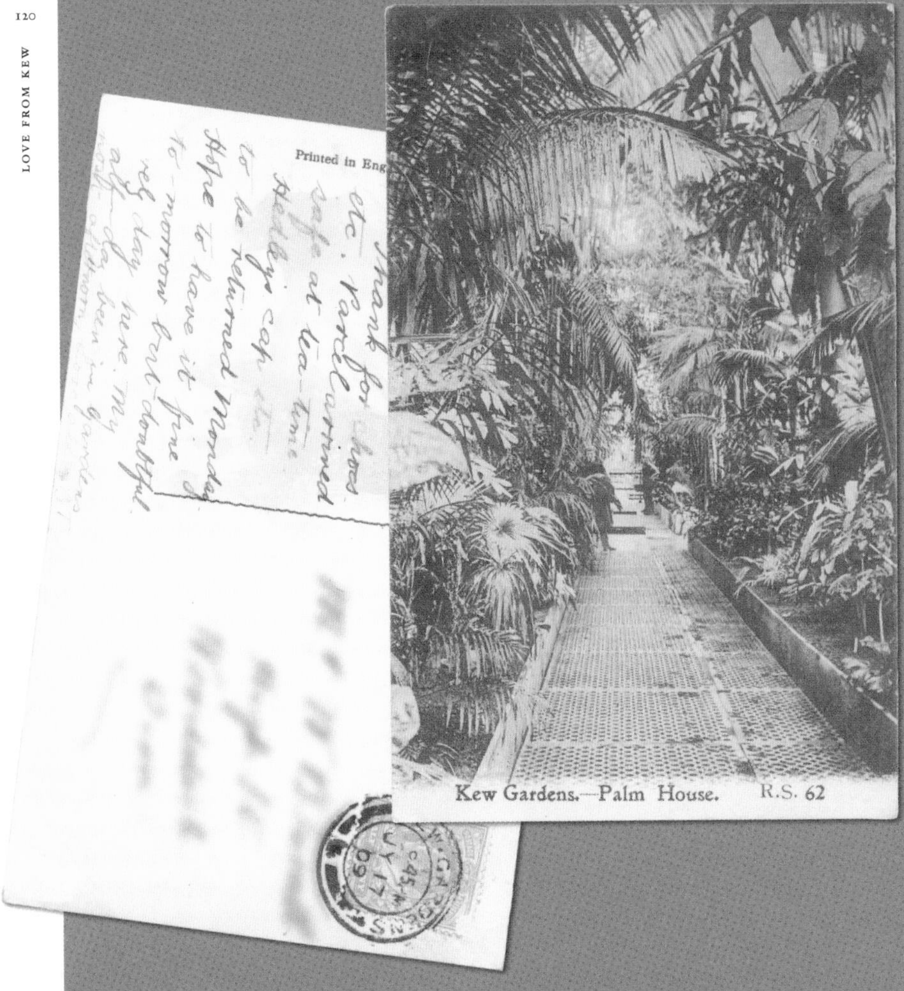

People move slowly through the garden in the heat of the afternoon. In the Palm House, mechanical misters hiss hot vapour onto shiny drip tips and the beehive ginger shakes its rattles at teak tubs mottled with moss. Two trapped robins sing their throaty warble from the stalks of a mangosteen, a big bass bottle palm joins the chorus.

'Lovely day here. My half-day. Been in Gardens most afternoon.'

This postcard showing the interior of the Palm House bears a Kew Gardens postmark and was sent to Woodstock in July 1909. Text printed on the left-hand side of the reverse reads: 'This space may be used for communication for Inland Postage and Foreign Countries, except Japan, Spain, and United States.'

THE PALM HOUSE, KEW GARDENS.

Plants push against the windows, tapping on the glass with broad waxy leaves. The air is filled with the scent of rare blooms. Exotic fruit burdens branches – yam, mango, star fruit, pip-filled pawpaw. Raffia reaches out her frondy fingers to hold hands with a dangle of bananas while rattans spear their way towards the canopy, fighting with the strangling tendrils of a knot vine.

The Temple of Bellona and the Pagoda can be seen in the background of this early postcard of the Palm House.

He shakes his umbrella and places it in the stand by the door of the Winter Garden. Twisting spiral staircases turn in the corners like wild vines snaking up the trunk of a tree. Water gurgles over the stones in the artificial falls and drips from mossy overhangs, pouring into pools as it winds through the gulley floor. Tree ferns grow on the banks and a lime parakeet flies down from a rafter, squawking from the leaves, picking orange berries with its beak.

'The flower beds were beautiful. I don't think you would like to have left them. It rained in Kew Gardens. London's a wonderful place.'

This postcard sent in July 1923 from Fulham to Bristol is labelled 'The Temperate House, or Winter Garden'. Winter Gardens are large conservatories built to house plants that will not survive outside during colder months.

THE VICTORIA GATES,
KEW GARDENS.

Time to go. She buys postcards from the kiosk at the
gate she shares a name with, feeling the sun glowing red
as it leaves the hemisphere, angry at its own departure.

This postcard sent from
Hampstead to Ealing shows
Victoria Gate which opened
in May 1989.

In the evening, the constable makes her rounds, scooping lost property into her sack. Reading glasses, bus passes, suncream, keys – slipped from pockets, lost from lives. On the Orangery terrace she finds a single starter sandal – size four, scuffed red leather and a T-bar strap.

The Orangery was built in 1757 as a home for orange trees, but the light levels inside were too low for them to flourish, and they were eventually transferred to the Kensington Palace Orangery in 1841. The building was then converted into a timber museum. In 1924 a kiosk was built next to the Orangery for the sale of guides and postcards. The building became a tea room in July 1989.

She walks in the garden,
an Eden on the river.
Nature's canvas, daubed
with purple and orange,
with frilled flowers, fruit,
leafy herbs, with pavilions,
temples, alcoves and urns
decorated with figs. A
citadel, a fortress, a walled
paradise where things grow
and blossom. Heaven, in an
ancient valley.

'Délicieuse promenade - Je suis
enchantée de Kew Gardens.
Affectueuses pensées.'

['Lovely walk – Kew Gardens was
delightful. With all my loving
thoughts.']

This postcard showing
Greenhouse Number Four and
bearing a Streatham postmark
was sent to Nogent-sur-Marne, an
eastern suburb of Paris, on 29 May
1921. It arrived two days later.

BIBLIOGRAPHY

Atkins, G., *Come Home at Once – Intriguing Messages from the Golden Age of Postcards*, Bantam Press, London, 2014.

Byatt, A., *Collecting Picture Postcards – An Introduction*, Golden Age Postcard Books, Malvern, 1982.

Carline, R., *Pictures in The Post – The Story of the Picture Postcard*, The Gordon Fraser Gallery Ltd, Bedford, 1959.

Desmond, R., *The History of the Royal Botanic Gardens, Kew*, 2nd edn., Royal Botanic Gardens, Kew, 2007.

Duval, W. and Monahan, V., *Collecting Postcards*, Blandford Press Ltd, Poole, 1978.

Evans, E. and Richards, J., *A Social History of Britain in Postcards – 1870–1930*, Longman Group Ltd, London, 1980.

Hill, C., *Picture Postcards*, 2nd edn., Shire Publications Ltd, Princes Risborough, 1991.

Holt, T. and Holt, V., *Picture Postcards of the Golden Age – A Collector's Guide*, MacGibbon and Kee Ltd, London, 1971.

Klamkin, M., *Picture Postcards*, David & Charles (Holdings) Ltd, Newton Abbot, 1974.

Lewis, G., *Postcards from Kew*, Her Majesty's Stationery Office, London, 1989.

Parker, L. and Ross-Jones, K., *The Story of Kew Gardens in Photographs*, Arcturus Publishing Ltd, London, 2013.

Price, K., *Kew Guide*, 2nd edn., Royal Botanic Gardens, Kew, 2017.

Skipwith, P. and Webb, B., *Edward Bawden's Kew Gardens*, V&A Publishing, London, 2014.

Staff, F., *The Picture Postcard and its Origins*, 2nd edn., Lutterworth Press, London, 1979.

ACKNOWLEDGEMENTS

My first and most profound thank you goes to a disparate group of complete strangers: the postcard writers who have afforded me a glimpse into a past I can never visit, no matter how many coins I push into Kew's old turnstile. I hope all your lives were well-lived.

I would also like to thank all the people associated with Kew who have told me fabulous tales about the landscape that captured my attention and my imagination.

I am enormously grateful to Gina Fullerlove and Lydia White at Kew Publishing, who understood exactly what I was trying to do and encouraged me with kindness and proper advice. I feel very privileged indeed – thank you. I would also like to thank Paul Little for digitising all the images I was lucky enough to find; Kiri Ross-Jones and the rest of the Kew's Herbarium, Library, Art & Archives team for dragging box after box of postcards in and out of the store cupboard; and Pei Chu, Chris Beard and Jo Pillai for all their help too. Thank you Ocky Murray who – I hope you will agree – has done a very fine job designing this book. Very many thanks to Michelle Payne, who found all the hideous errors I made late at night. *Merci beaucoup* to Aline Dufat for her kind help with French translation. Thank you Katie Read from Read Media for some excellent ideas.

Thank you to the staff at The Postal Museum who helped me to find the books I was looking for.

I am very grateful to Dame Judi Dench, Tracy Chevalier, Francis Pryor and Lisa Woollett for saying such lovely things about my writing.

Thank you, Mum, for lots of things, but specifically here for your help with understanding pre-decimalisation!

DB – thank you for your humour and patience while you watched me burn through my latest obsession and fill our home with yet more musty things from the past.

ABOUT THE AUTHOR AND BOOK

Postcards let us look into lives – their scribbled stories are cryptic, funny and poignant. In *Love From Kew* the everyday is captured in inky messages about birthdays, snow, exams, dancing, and love, written on picture postcards from Kew Gardens. Told with wonder and imagination, author Sophie Shillito entwines these tiny tales with secrets from the Kew landscape which mumble through the turf and whisper in the bushes.

A postbag full of scribbled stories from the last century, *Love From Kew* sings with the spirit of the Gardens. It offers a fascinating insight into the social anthropology and micro-history of the past 100 plus years, revealing the human need to connect with Kew and the natural world, as well as the importance of communicating with loved ones – two themes that are ever more precious and valued in recent times.

Sophie Shillito writes about place, palimpsests, and micro-histories. Her first book, *All The Little Places* (Blue Mark Books, 2020) is a haunting poetic-prose tapestry of fairytale fragments. Sophie lives in London. She likes mudlarking and mountains.

www.sophieshillito.com